LET'S INVESTIGATE
Graphs

LET'S INVESTIGATE
Graphs

By Marion Smoothey
Illustrated by Ann Baum

MARSHALL CAVENDISH
NEW YORK • LONDON • TORONTO • SYDNEY

© Marshall Cavendish Corporation 1995

Published by Marshall Cavendish Corporation
2415 Jerusalem Avenue
PO Box 587
North Bellmore
New York 11710

Series created by Graham Beehag Books

Editorial consultant: Prof. Sonia Helton
University of South Florida, St. Petersburg

Library of Congress Cataloging-in-Publication Data

Smoothey, Marion,
      Graphs / by Marion Smoothey : illustrated by Ann Baum. – Library ed.
        p. cm. – (Lets Investigate)
      Includes index.
      ISBN 1-85435-775-1 ISBN 1-854535-773-5 (set)
      1. Graphic methods – Juvenile literature. [1. Graphic methods.]
      I. Baum, Ann. ill. II. Title. III. Series: Smoothey, Marion, 1943-
      Lets Investigate.
      QA490.S58 1995   94-13133
      001.4'226 – dc20   CIP
                                      AC

Printed in Malaysia by Times Offset (M) SDN BHD

Contents

Introduction

6

A graph can be a useful way of displaying information.
It can also help you to solve problems and make
predictions. This book shows you how to understand and
draw graphs. You will need graph paper and paper
with a square dotted grid for drawing your own graphs.
You may copy pages 59 and 60 for this. You will also
need tracing paper, a ruler, an eraser, and a sharp pencil.

Conversion Graphs

Two main systems of measurement are used around the world. They are called the metric system and the imperial system.

The metric system measures distances in millimeters(mm), centimeters (cm), meters (m), and kilometers (km).

$$10 \text{ mm} = 1 \text{ cm}$$
$$100 \text{ cm} = 1 \text{ m}$$
$$1{,}000 \text{ m} = 1 \text{ km}$$

The imperial system uses inches (in. or ″), feet (ft. or ′), yards (yd.), and miles (mi.).

$$12 \text{ in.} = 1 \text{ ft.}$$
$$3 \text{ ft.} = 1 \text{ yd.}$$
$$1{,}760 \text{ yds.} = 1 \text{ mi.}$$

A foot was originally the length of a person's foot, and a mile was 1,000 paces of a Roman **legionary**.

A graph can be a useful way of converting one set of measurements to another. You can use this graph to change miles into kilometers and vice versa.

Conversion Chart for Miles/Kilometers

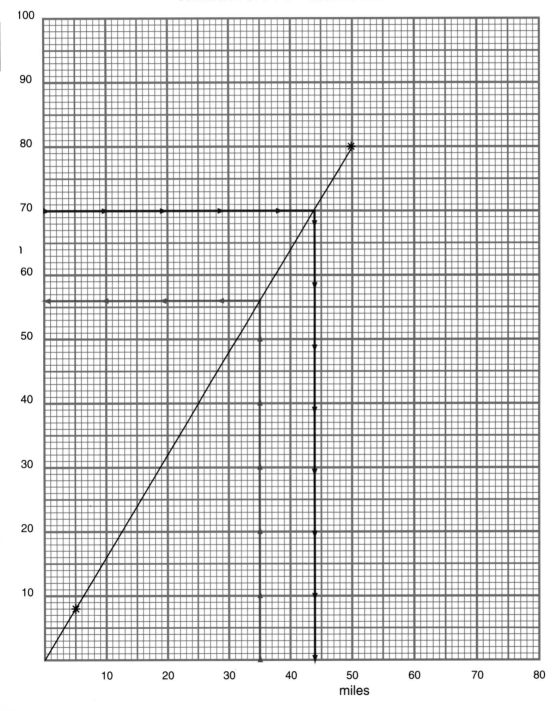

8

Tony's dad has run 35 miles in a charity run, and Tony wants to write and tell his grandmother, who lives in Italy, about it. She is used to figuring in kilometers, and has promised to pay 25 cents a kilometer. Tony wants to figure out how much she owes his dad.

The red line shows you how to convert 35 miles into kilometers. First find 35 along the mile scale, which in this case is the **horizontal** scale. Use a ruler to draw a line up to the sloping graph line. Then draw a straight line to the kilometer scale and read the result. It is 56.

The conversion graph opposite gives you the information that 35 miles = 56 kilometers. This is not exactly true. A closer measurement, 35 miles = 56.3255 kilometers, can be found by using a calculator but the graphical answer is good enough for Tony.

● **1.** How much does Tony's grandmother owe his dad?

You can use the same graph to change kilometers into miles. The blue line shows you how to change 70 kilometers into miles. Draw a straight line from 70 on the kilometer scale to the graph line. Then draw a straight line down to the mile scale and read the result.

● **2.** How many miles approximately equal 70 kilometers?

Travel Graphs

When you take a trip, you travel a distance in a period of time. This can be shown as a graph. Usually the time is shown on the **horizontal axis** and the distance on the **vertical axis**.

10

This graph shows a journey that could fit this story. Sean ran to the store to buy a doughnut (the red part). He began to feel out of breath and slowed down (the blue part). When he arrived at the store he stopped to look in the window (the green part).

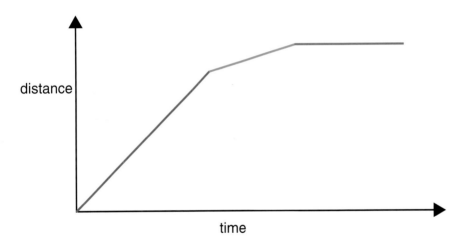

Notice that the faster you travel, the more distance you cover in a given time. The graph line of your journey slopes steeply when you are traveling quickly.

● **1.** What happens to the graph line when you are standing still?

● Match the stories to the correct graphs.

A. Rosa set off for school. As she was walking along, she met her aunt with her dog and stopped to play with it. Rosa realized she was going to be late for school and ran the rest of the way.

B. Rosa set off for school. After a while she realized she had forgotten her homework and ran back to fetch it.

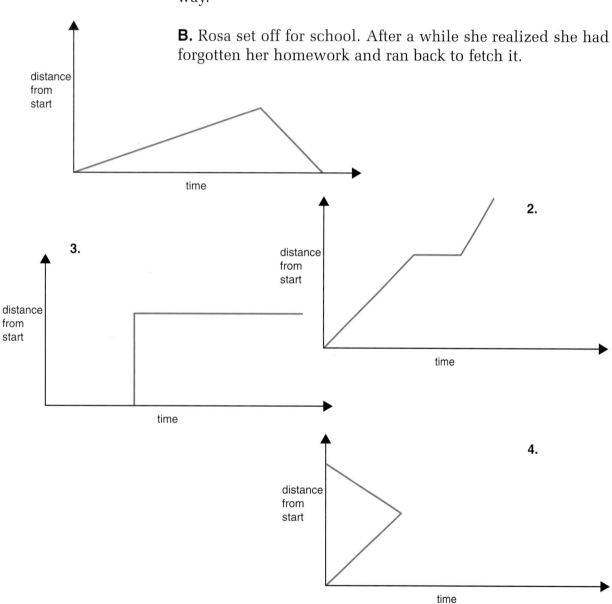

Two of the graphs cannot show journeys. Explain what is wrong with them.

Understanding Graphs

12

● Match the drawings to the graphs.

A. weight / weeks

B. height / months

C. height / years   1993

Bath Time

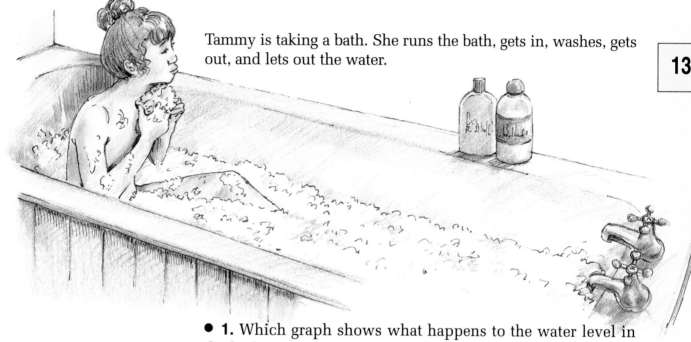

Tammy is taking a bath. She runs the bath, gets in, washes, gets out, and lets out the water.

● **1.** Which graph shows what happens to the water level in the bath?

● **2.** What is wrong with each of the other graphs?

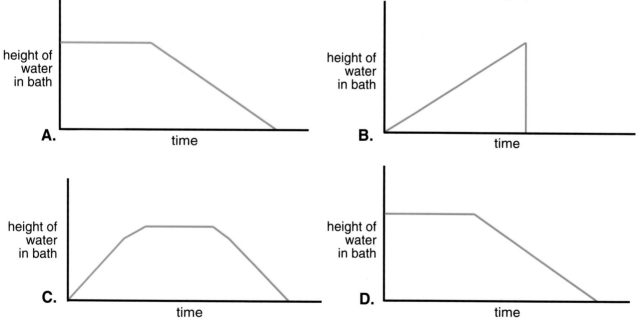

Answers to page 13

C. shows the water level correctly.

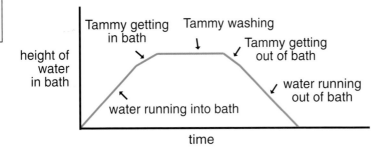

A. shows the bath with water in it already. It does not show the bath filling. There is no change of level as Tammy gets in and out.

B. shows the bath filling up, and then the water running out of it instantly, which is impossible. The water will take some time to flow down the drain.

D. does not show the change in water height as Tammy gets in and out of the bath.

Archimedes discovered his theory of water displacement while in his bath!

The height does not have to be shown on the **vertical axis**. You can use the **horizontal axis** for it.

- Which graph shows Tammy taking a bath now?

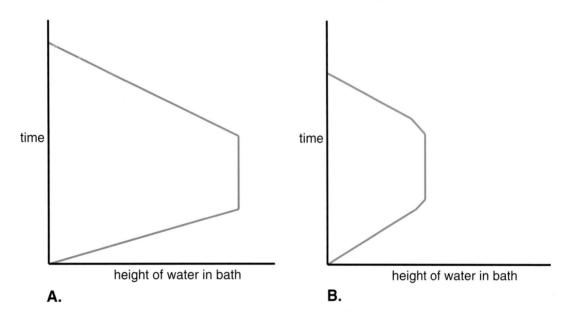

A.

B.

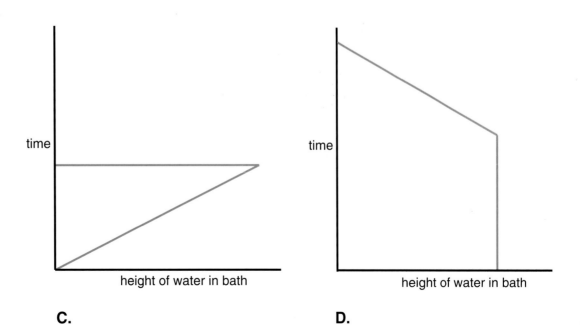

C.

D.

Up in the Air

Not all graphs are straight lines. Some are curves.

This graph shows a ball being thrown.

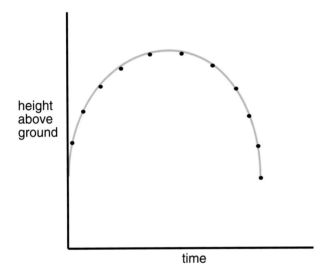

A graph is NOT a picture. It is a way of representing two pieces of information in one place, and sometimes joining these pieces of information to make a line.

If you look at the shape of the line without thinking about the information it is giving you, you might think it shows a ball being thrown from one person to another. It doesn't!

Look again. The curved line is made up of many points. A few of the points are shown. The horizontal axis is telling you about the ball at a given *time*. The vertical axis is giving you information about the ball's *height*. Each point on the line tells you the height of the ball above the ground at a particular time.

You can see that the ball gets higher and higher above the ground for a period of time and then gets lower and lower until it comes back to the height it started at. The graph is showing you what happens when a ball is thrown *up in the air*.

The same information is shown on this graph.

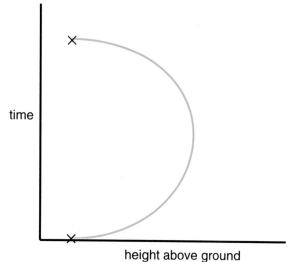

The curved line shows that the ball begins at a certain height above the ground, then for a period of time the ball travels further and further from the ground, and then it falls nearer and nearer to the ground until it reaches the height from which it started.

Notice that the graphs do not start at a zero height.

● Why is this?

Using Scales

Except for the conversion graph at the beginning, the rest of the graphs you have looked at so far in this book have given you some information but have not been very precise. This is because the graphs have not had scales marked along the axes. (Axes is the word used for more than one axis. It is pronounced *ak' sees*).

The graphs on pages 16 and 17 do not tell you how high the ball was thrown or how long it was in the air. The graph of Tammy's bath does not tell you how deep the water was or how long she stayed in the bath.

Look at this graph recording the temperatures at noon in Cactus Canyon during one week.

Temperature at Noon in Cactus Canyon

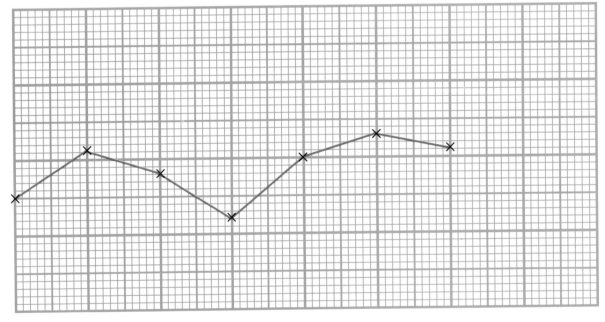

● **1.** On what day was this picture drawn?

● **2.** What was the hottest temperature recorded?

● **3.** What was the difference between the highest and lowest temperatures?

The questions on page 19 are impossible to answer with the graph given, because it has no scales. You need a graph like this one.

20

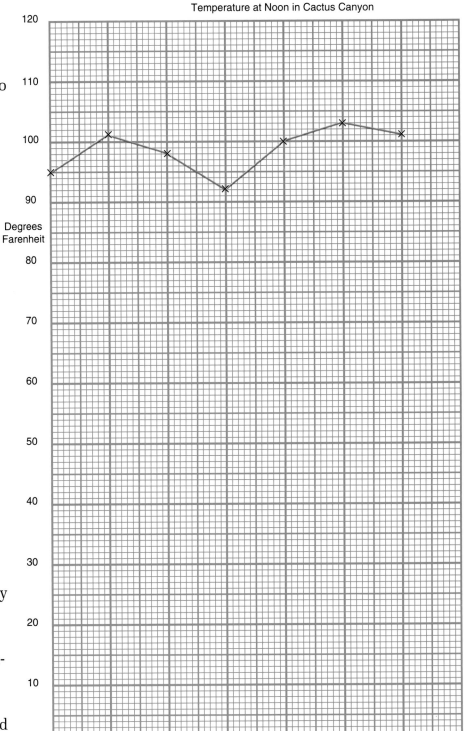

Temperature at Noon in Cactus Canyon

Degrees
Farenheit

MON   TUES   WED   THURS   FRI   SAT   SUN

Now you can see that:

1. The picture was drawn on Monday because that is the only day that the temperature was 95° at noon.

2. The hottest temperature was 103°.

3. The difference between the highest and lowest temperatures is 103°– 92° = 11°.

It is most important that the scale remains the same all the way along the axis. For the Cactus Canyon graph, ten small squares on the horizontal axis represent 24 hours.

● **1.** What does one small square represent on the vertical axis?

The space at the bottom of the graph of temperatures in Cactus Canyon is wasted because the lowest temperature recorded at noon was 93°. The part of the scale from 0 to 90 is not needed, but one square must stand for the same number of degrees Fahrenheit all the way up the scale.

There is a way to overcome this problem. You can use a zigzag to show that you have omitted the part of the scale that is not needed.

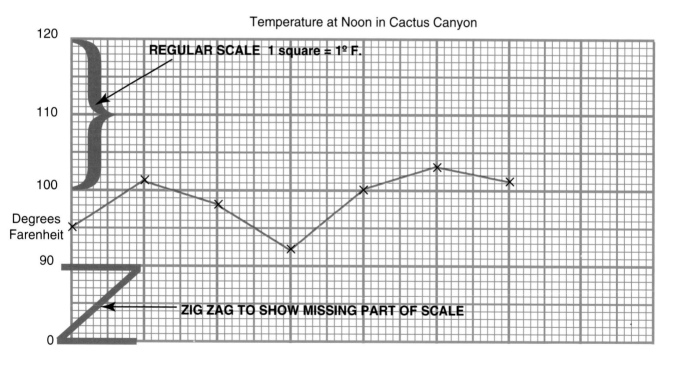

Temperature at Noon in Cactus Canyon

For the rest of the axis, the scale must remain the same all the way along. On this graph, one small square stands for 1°F. You can use any scale you like as long as it will fit into the space you have available.

At first glance, this graph looks different from the one on page 21, but in fact, it gives exactly the same information and is equally correct. Different scales have been used so that the variations in temperature from day to day are more noticeable.

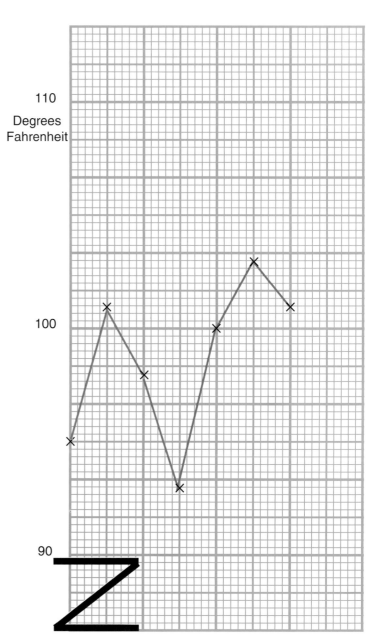

Check that the scales are regular all the way along each axis.

● **1.** How many squares equal 24 hours on the horizontal axis?

● **2.** How many squares equal 1° F on the vertical axis?

Check that the temperatures recorded each day are the same in both graphs.

Choosing a Scale

When you choose a scale, you must make sure it will fit onto the paper you have available. Mark the scale lightly in pencil first, so you can avoid running the graph off the edges of the page.

It is a good idea to choose scales that are easy to work with. Usually it is best to work in groups of 1, 5, or 10 small squares, because they are easy to count.

You must label your scales and give your graph a title. On the Cactus Canyon graphs, it was necessary to explain that the numbers on the vertical axis were degrees Fahrenheit and that the temperatures were taken at noon.

Without these two pieces of information, you might have thought the graph was about the number of people buying a doughnut at the local store or something similar.

Drawing a Graph to Convert Grams to Ounces

24

Find out which of these products gives you more detergent for your money by drawing a conversion graph for grams and ounces.

Make a copy of the graph paper on page 60.

Some information to help you

16 oz. = 1 lb.
1,000 gm = 1 kg

1 ounce = 28 grams (to the nearest whole number)
10 oz. = 280 gm
20 oz. = 560 gm

Use a scale of 2 small squares = 1 ounce for the horizontal scale and 1 small square = 10 grams for the vertical. This will be easy to count and will make the best use of the size of the paper.

Mark in the positions for 10 oz. = 280 gm and 20 oz. = 560 gm.

Draw a line through these two pairs of points and through 0.

Now draw lines on your graph to see which packet has more laundry detergent.

If you have forgotten what to do, turn back to page 9

● **1.** Which package gives you more laundry detergent?

Use your graph to make some price comparisons of your own between real products.

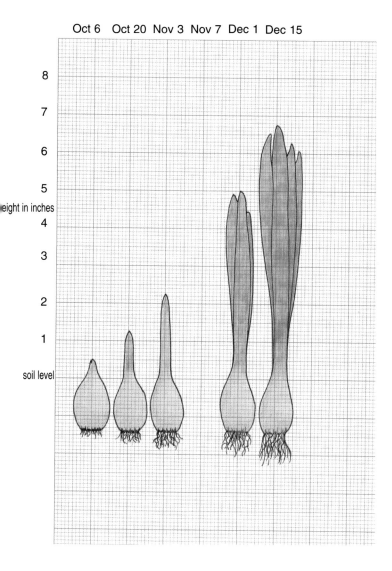

Oct 6   Oct 20 Nov 3 Nov 7 Dec 1 Dec 15

height in inches

8
7
6
5
4
3
2
1
soil level

Straight Lines and Curves

Tim planted a hyacinth bulb in a bowl on Sunday, September 22 as a Christmas present for his mom. He measured the height of the bulb's growing tip above the soil level each alternate Sunday for twelve weeks, except for November 17 when he forgot. His results are recorded on this diagram.

The measurements can be recorded in a table like this.

DATE	Oct. 6	Oct. 20	Nov. 3	Nov. 17	Dec. 1	Dec. 15
HEIGHT	$\frac{1}{2}'$	$1\frac{1}{4}''$	$2\frac{1}{4}''$		$5''$	$6\frac{3}{4}''$

Turn to page 27.

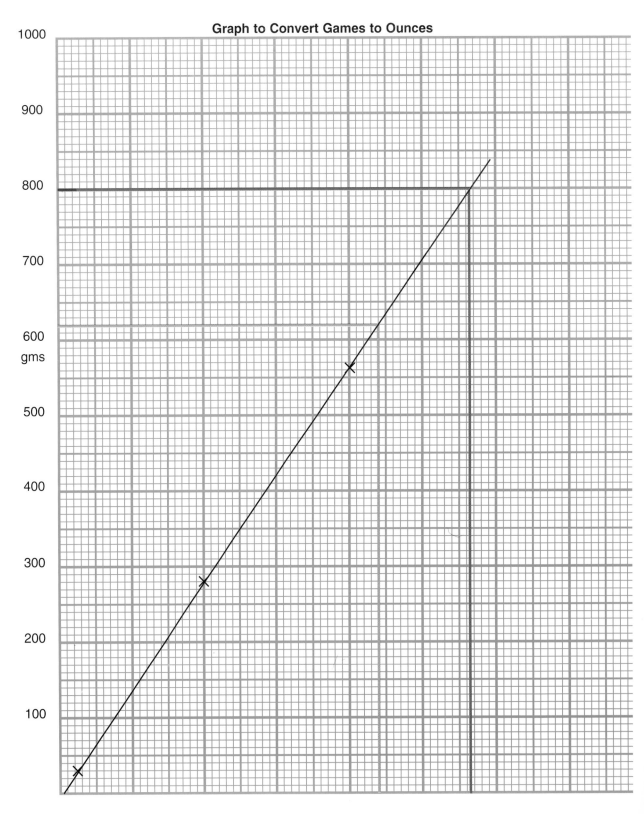

Graph to Convert Games to Ounces

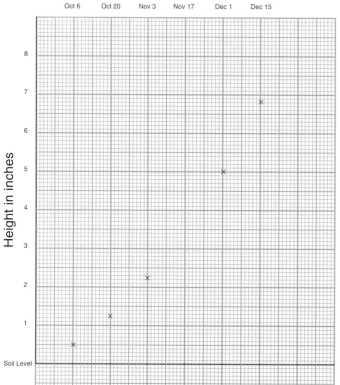

The information in the table on page 25 can be represented by points on a graph like this.

Notice that the scales are regular on each axis, the axes are labeled, and the graph has a title.

To complete the graph, draw a line through the points. Place a ruler on the graph and see if there is any way of joining all the points with a straight line. The drawing shows a ruler joining two of the points but missing the other three points.

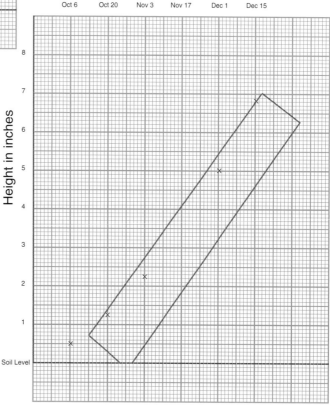

It is impossible to find a straight line to fit all the points. In this case, the line of the graph is a curve.

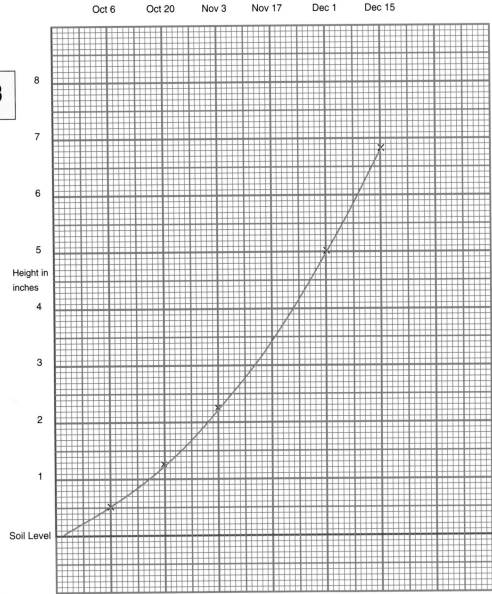

● **1.** Use the graph to find out the height of the bulb's growing tip above the soil on November 17.

When you are drawing a curve, use a pencil and draw lightly so that you can erase it if necessary. You are trying to achieve a smooth curve. Do NOT use a ruler to draw a series of straight lines joined together.

28

Journey Times

The average speed for a journey is found by dividing the distance traveled by the time taken. The table below shows the times that the journey from Atlanta, Georgia to Jackson, Mississippi takes at different average speeds.

SPEED IN MPH	30	40	50	60	70
TIME IN HOURS	12·7	9·5	7·6	6·3	5·4

Draw a graph of these results. Choose reasonable scales, and use the information in the table to plot five points. Remember to label your axes and give your graph a title.

● **1**. Is the line you need to draw to join the points straight or a curve?

Use your graph to answer these questions.

● **2**. How long will the journey take at
(a) 45 mph, **(b)** 55 mph?

● **3**. If you have to be in Jackson by noon and estimate that you will be able to average 35 mph, what time must you leave Atlanta?

Scatter Graphs

In some graphs, it does not make any sense to join the points to make a line, whether straight or curved.

Aaron recorded the waist and height measurements of himself (**A**), Brian (**B**), Charles (**C**), Diana (**D**), Emily (**E**), Fran (**F**), Gerald (**G**), and Hannah (**H**) on this graph.

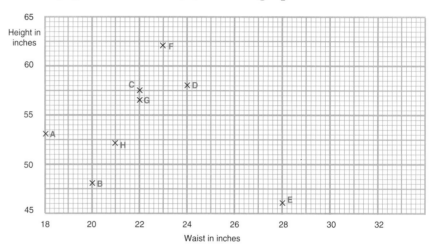

These are separate pairs of measurements. Each cross on the graph represents a person and gives you information about his or her body shape. Each cross tells you one person's height and waist measurement in inches.

● **1.** Can you identify the people in this drawing? Gerald is the clue.

Gerald

It does not make sense to try and join the points together to make a line, because each point is a separate person. We can, however, see that most of the points are clustered near a diagonal line that we can draw on the graph.

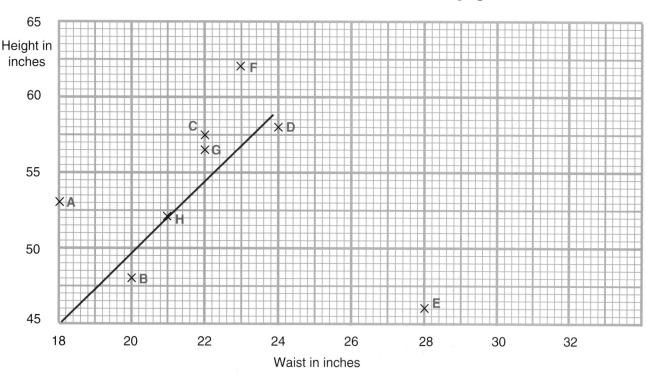

This tells us that most of the people represented on the graph have height and weight measurements in approximately the same proportion. Although their sizes vary, their body shapes are similar. The crosses on the graph that are not near the line represent the people who would be more noticeable in the group because their body shapes are different from the rest.

● **1.** Patsy is of average build and her waist measures 23″. About how tall is she?

Aaron then asked the same people how many pets they had, and drew another scatter graph of the results.

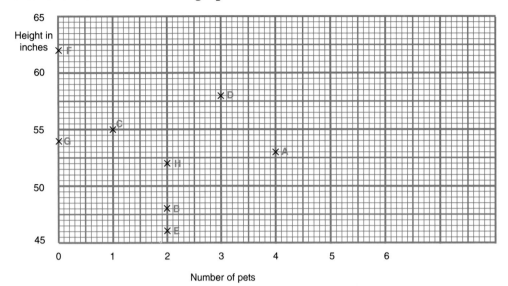

Again, it makes no sense to join the points to make a straight line or a curve.

● **1.** What information is each cross on the graph giving you?

● **2.** How many pets does Gerald have?

This graph is different from the height/waist measurement graph because the crosses do not cluster along a diagonal line. This is telling you that people's heights are no guide to how many pets they have. There is no **correlation** between height and number of pets.

The crosses clustering near the diagonal line on the height/waist measurement graph show you that there is some correlation between a person's height and his or her waist measurement.

This means that generally, a tall person will have a larger waist then a short person. This is only a general rule; you can find cases where it is not so. The crosses on the graph that are not near the diagonal line represent some of these people who are exceptions to the rule.

Conduct some surveys of your own by asking for two pieces of information about each person. Record the results on a scatter graph to see if there is any correlation.

Clothes manufacturers use the general rule to design their garments.

32

Ordered Pairs

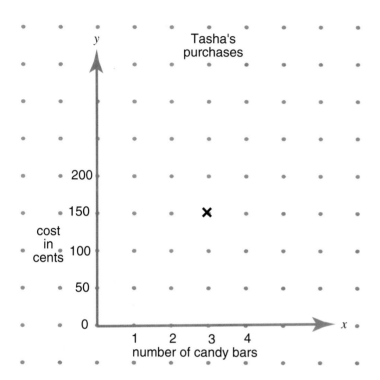

A graph is a way of recording two pieces of information at one point. This is done by using one axis for one piece of information and a second axis for the second piece of information. A mark is made where the two pieces of information meet.

In this example, the two pieces of information are that Tasha bought three candy bars and that she spent 150 cents. They are represented by the cross on the graph.

Notice that the axes have regular scales, they are labeled, and the graph has a title. The axes are also labeled x and y.

The information to be put on a graph is usually presented in a coded form like this (x, y). You need to be able to crack the code. The brackets and the comma between the two numbers is a clue that these numbers need to be related to a pair of axes.

The next important thing is to know which number refers to the horizontal, or x axis, and which refers to the vertical, or y axis. Mathematicians have agreed that the first number tells you how far to go along the x axis, and the second number is your instruction for the y axis.

In the example of Tasha's candy bars, the information is coded (3, 150). The cross is to be placed 3 across and 150 up.

It is important that you remember that the FIRST number tells you how far ACROSS to go and the SECOND tells you how far UP to go. The ORDER of the pair of numbers is important.

One way of remembering this is to remember that A, the first letter of ACROSS, comes before U, the first letter of UP. Another way is to remember "ACROSS the hall before UP the stairs."

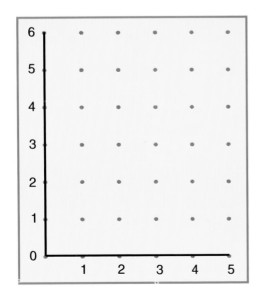

Ivory

This is a game for two players. You each need a pencil and paper. Imagine one of you is the game warden of Bukhiva National Park – a ficticious wildlife park in Africa – who is having problems with ivory poachers. You have three small herds of elephant in the park which you are trying to protect. The other player is an ivory trader who has three hunters hidden in the park.

Each of you draws axes and scales like these on your piece of paper.

The game warden plots the position of the three herds on the piece of paper and keeps it hidden.

In this example, the ordered pairs are (1, 2), (3, 4), and (4, 6).

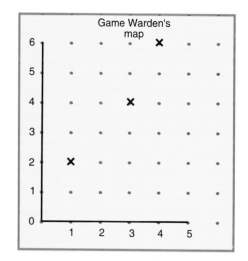

35

The poacher plots the positions of the three hunters on the graph and keeps it hidden. This poacher has hidden hunters at (2, 2), (4, 1), and (5, 3).

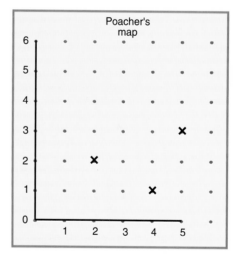

Round 1

The game warden guesses the position of a hunter by giving the ordered pair (3, 3), and marks it in a different color from the herds on his copy of the graph.

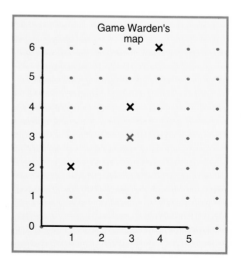

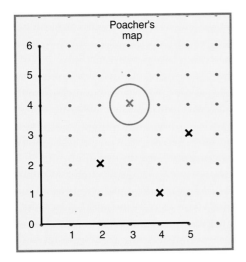

The poacher tells the game warden he has not scored a hit. He guesses the position of a herd by giving an ordered pair. In this example he chooses (3, 4). He marks (3, 4) on his copy of the graph in a different color from the hunter crosses.

The game warden tells the poacher this is a hit. The poacher circles his colored cross to record a hit.

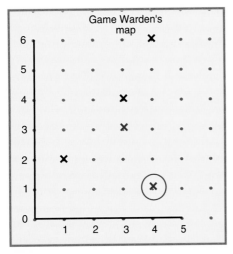

Round 2

The game warden tries (4, 1) and the poacher tells him he has scored a hit.

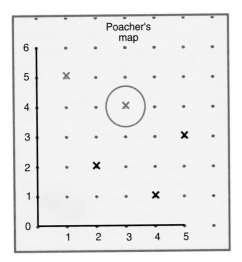

The poacher chooses (1, 5) which is a miss.

Round 3

The game warden scores another miss with (1,3)

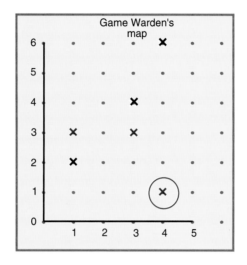

The poacher also misses with (4, 2).

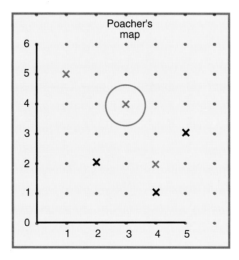

The game continues until one player scores three hits and wins the game.

Games with Dice

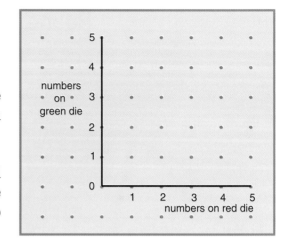

You need two dice of different colors. Cover the six dot face of each of them with a piece of a stick-on label and call those faces 0.

Make a copy of the grid (right). This is for a red and a green die. Write whatever colors your dice are. Each player chooses a different color to record his/her hits.

Rules of the game

Take turns throwing both dice. Add the scores on the two dice together.

If they total 6, you score a hit. Mark a cross in the correct place on the grid.

If the two numbers do not total 6, you do not score a hit and you do not mark the grid.

The winner is the first person to score three hits.

The graphs (right and on the next page) show you the results of a game between Bob and Ruth. Bob has recorded his hits in purple and Ruth has used orange.

Game 1 – Round 1

Ruth's throw (3, 4) – not a hit (3 + 4 does not = 6).

Bob's throw (2, 4) – a hit (2 = 4 = 6). Bob records his hit on the graph.

The score looks like this:

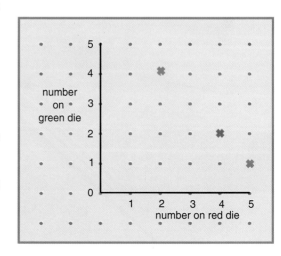

38

Round 2

Ruth's throw (0, 5) – not a hit
(0 + 5 does not = 6).

Bob's throw (3, 1) – not a hit.

The score looks the same as after round 1.

Round 3

This is the score after round 3.

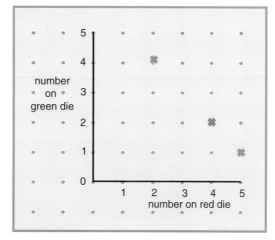

● **1.** What number did Ruth throw on the green die?

● **2.** Is this what Bob threw?

Round 4

Ruth threw a double 3 (3, 3). Bob threw (4, 2). Ruth had already scored a hit on (4, 2), so at the end of the round, the score looked like this:

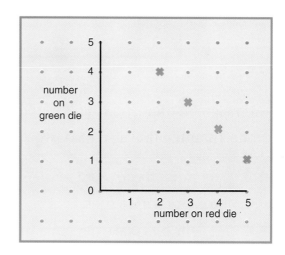

Round 5

This is the score after round **5**.

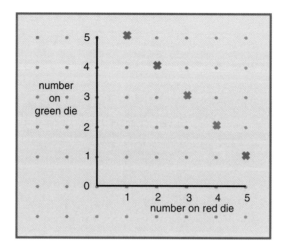

40

● **1.** Who has won?

● **2.** What was the winning throw?

The crosses on the graph all lie on a straight line.

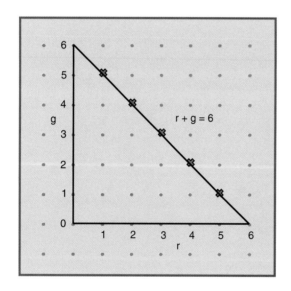

If we use r for ``number on the red die'' and g for ``number on the green die,'' the line joins the points where r + g = 6.

Notice that the line also passes through (6, 0) and (0, 6).

● **3.** Why don't these score in the game?

Play the game with a friend.

● **4.** Did your hits make the same line as the one on this page?

Play the game again, but this time the dice have to total 4 for a hit.

This is the line for r + g = 4.

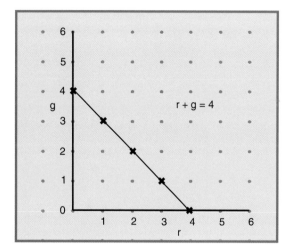

r + g = 4

● **1.** How many possible winning throws are there in this game?

● **2.** Are your hits on the line r + g = 4?

Now play the game so that you have to throw a double to score a hit.

● **3.** What are the possible throws for scoring a hit when you have to throw a double? Write them as ordered pairs in the form (r, g).

A double means "the number on the red die = the number on the green die," or r = g.

r = g is an example of an **equation** because it is a statement about one thing equaling another.

● **4.** What was the equation in the first game (pages 38-41)?

● **5.** Draw the line for r = g on a graph. Remember to label your axes.

● **6.** Write down as ordered pairs the numbers on the two dice of throws that are hits on the line r = g.

● **7.** Check that these ordered pairs are the same as your answers to question 1.

Look carefully at this graph. The line joins the scores for hits in another game with two dice.

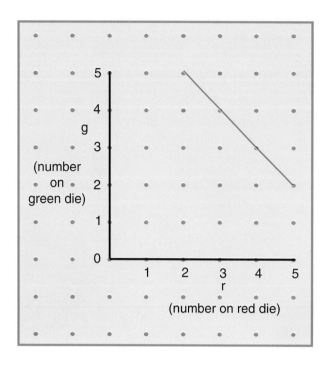

● **1.** How many throws are there that can score a hit in this game?

● **2.** Write down the throws that score a hit in this game as ordered pairs in the form (r, g).

● **3.** What equation fits the points on the line?

● **4.** Why might this game end in a draw?

Coordinates with Negative Numbers

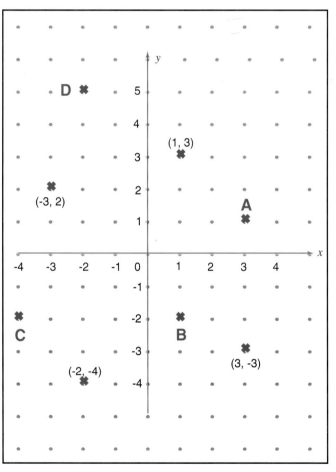

Another name for ordered pairs is **coordinates**. If we extend the axes, it is possible to use negative as well as positive coordinates.

So far we have looked only at the numbers greater than 0 on a pair of axes. If necessary, we can extend the axes backwards and downward to show numbers less then 0. These numbers are called **negative numbers**.

The numbers to the left of 0 on the x axis and below 0 on the y axis are less than 0, or negative numbers. In the examples on this and the next page, the axes stretch from −4 to 4 on the x axis and from 5 to −5 on the y axis. When you draw a graph, you can make the axes stretch as far as you wish.

Look carefully at the coordinates given for the four marked points. Follow the rule for coordinates of ACROSS then UP (or down).

● What are the coordinates for the points **A**, **B**, **C**, **D**?

44

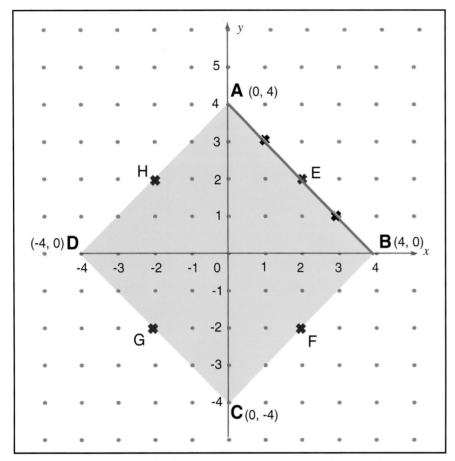

The coordinates of the corners of the yellow shape are labeled for you. Notice again that the order is still the ACROSS number first, then the UP (or down) number.

● **1.** What is the name of the yellow shape?

● **2.** What are the coordinates of the mid-points of each side? (They are each marked with a purple cross and labeled **E, F, G, H.**)

● **3.** What are the coordinates of the other two points marked on the blue line?

● **4.** What is the equation of the blue line?

☆ **Hint**
You should have an "*x*," a "*y*," a "+," a number, and an "=" in your answer.

Fraction and Decimal Coordinates

45

It is possible to have coordinates that use fractions or decimals.

The coordinates of points **A** (½, ½) or (0.5, 0.5), **B** (−1½, −½) ór (−1.5, −0.5), and **C** (0, −1¼) or (0, −1.25) are given.

● **1.** What are the coordinates of points **D** and **E** using fractions or decimals?

> **Check your answers on page 63 before continuing.**

Coordinates Check-up

Make a copy of the grid of squared dots on page 59. Use your copy to draw a pair of axes extending from −5 to +5.

Plot these points and join each point to the next with a straight line.

(3, 2), (3½, 1), (5, 0), (4½, −1), (3, −½), (2¼, −2), (2½, −3½), (1¾, −3½), (1½, −2)

Start again: (1½, −3), (1, −3), (1, −2), (−2½, −1¾), (−2½, −3¼), (−3½, −3¼), (−3½, −2), (−5, 1), (−4,½), (2½, 1), (3, 2)

Start again: (−3½, −3), (−4, −3), (−4, −1). Draw a small circle at (3½, ½) and then finish the drawing as you wish.

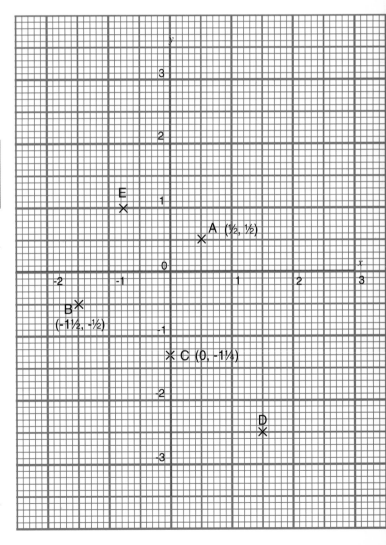

Families of Lines

46

In the middle of a page, draw axes from −6 to 6. Label them x and y and write in the scale. Use one square for one unit. Plot the points (3, 3) and (−3, −3). Draw the line that passes through the two points.

● **1.** What is the equation of the line? (What rule does every point on the line obey?)

> **Check your answer on page 53 before you continue.**

Now you are going to draw the line for the equation $y = x + 4$. That equation means that for every point on the line, the y coordinate of the point is 4 more than the x coordinate of the point. For example when $x = 2$, $y = 2 + 4 = 6$.

● **2.** Make a copy of this table, and fill in the missing numbers to find the coordinates of some of the points on the line $y = x + 4$.

x	$x + 4 = y$	(x, y)
2	2 + 4 = 6	(2, 6)
4	4 + 4 = 8	(4, 8)
6	6 + 4 = ?	(6, ?)
0	0 + 4 = ?	(0, ?)
−3	−3 + 4 = 1	(−3, 1)

● **3.** Plot any two of the points on your graph. Join them and draw the line $y = x + 4$.

● **4.** Make a similar table for $y = x - 4$ to find pairs of coordinates for which this equation is true. Draw the line for the equation.

> **Check your answers on page 54.**

Look carefully at the three lines you have drawn. They are **parallel** to each other.

The line $y = x + 4$ cuts the y axis at 4 and the x axis at -4. The line $y = x$ cuts the y axis at 0 and the x axis at 0.

● **1.** Where does the line $y = x - 4$ cut the y axis and the x axis?

● **2.** Think about the line $y = x + 2$. Will it be parallel to the three lines you have drawn? Where will it cut the y and x axes?

● **3.** Find the equations of some more lines parallel to $y = x$.

48

On another piece of dotted paper, draw a pair of axes from −8 to +8. Label them x and y. Draw in the scale as before.

● **1.** Make a table for each of these equations:

$y = 2x$
$y = 2x + 4$
$y = 2x - 4$

Use the values $x = -2$, $x = 0$ and $x = +2$.

● **2.** Draw the lines for the equations on your graph and label them.

● **3.** Are the lines parallel to each other?

● **4.** Where do each of the lines cut the y and x axes?

Compare these three lines with the three lines you drew for $y = x$, $y = x + 4$, $y = x - 4$.

● **5.** In what ways are the two sets the same?

● **6.** In what way are they different?

Make a table for the equation $y = 5x$.
Draw the line for $y = 5x$.

● **7.** Is it parallel to the other three lines on your graph?

Families of Curves

On a new piece of dotted paper, draw and label an x axis from −4 to +4 and a y axis from −4 to +12.

● **1.** Make a table for $y = x^2$, with the values of x being −3, −2, −1, 0, 1, 2, and 3 (x^2 means $x \times x$).

When you multiply two negative numbers together, the product is always positive.

For example: $-2 \times 2 = 4$.

● **2.** Plot the coordinates from your table as points on your graph.

Join them to make a smooth curve.

● **3.** Make tables and draw graphs for $y = x^2 + 2$ and $y = x^2 - 3$.

Check your answers on pages 56 and 57.

Trace the curve for $y = x^2$ onto tracing paper.

Move the tracing up and down the y axis.

● **1.** What do you notice?

Fold the tracing in half vertically.

● **2.** What do you notice?

Make tables for $y = 3x^2$, $y = 3x^2 + 4$, and $y = 3x^2 - 5$.

On a fresh piece of dotted paper, draw suitable axes, plot the coordinates, and draw the curves.

Make a tracing of the curve $y = 3x^2$

● **3.** Does it fit the other two curves on page 58?

● **4.** Does it fit the three curves on page 57?

● **5.** What difference does the number before the x^2 make?

Gradients

The steepness of line is called its **gradient**.

You see gradients on traffic signs warning you when you are approaching a steep hill.

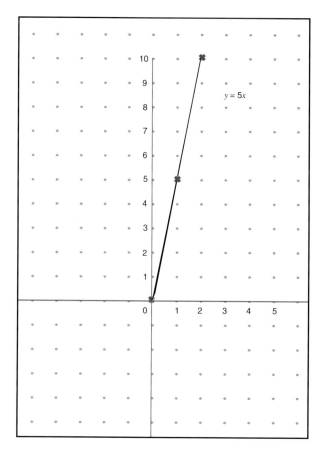

The part of the equation that gives you information about the gradient of a line is the number before the x. It tells you how far up the line moves as it moves across.

$y = x$ is the same as $y = 1x$: the gradient of the line is 1. The line moves up 1 as it moves across 1.

$y = 2x$ is telling you the line has a gradient of 2. The line moves up 2 as it moves across 1.

● **1.** What are the gradients of these lines:
(a) $y = 5x - 3$
(b) $y = \frac{1}{2}x$
(c) $y = x + 10$

The equation of a line gives you information about its gradient.

You have found that the lines $y = x$, $y = x + 4$, and $y = x - 4$ are parallel.

You have also found that the lines $y = 2x$, $y = 2x + 4$, and $y = 2x - 4$ are parallel.

51

$y = 5x$ is not parallel to either of the other two sets of lines you have drawn.

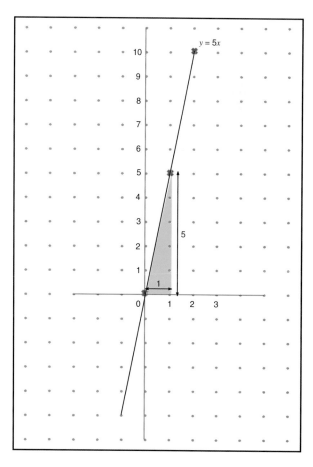

Answer to page 45

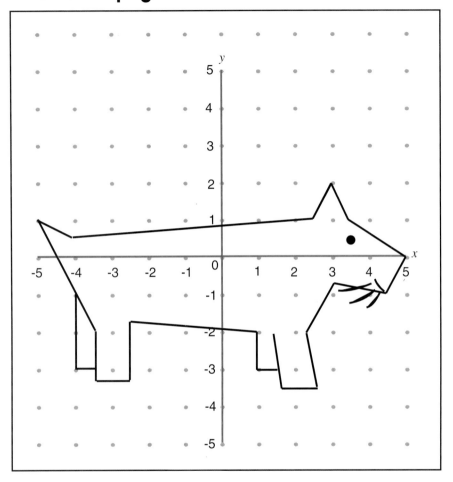

If your shape is not right, check carefully to see where you have gone wrong.

Perhaps you forgot to count ACROSS first, or perhaps you just did not count carefully enough, or perhaps you got confused counting fractions backward.

If you cannot see where you went wrong, ask someone to help you. Try making a coordinate puzzle like this for a friend to solve.

Turn back to page 45.

Answer to page 46, question 1

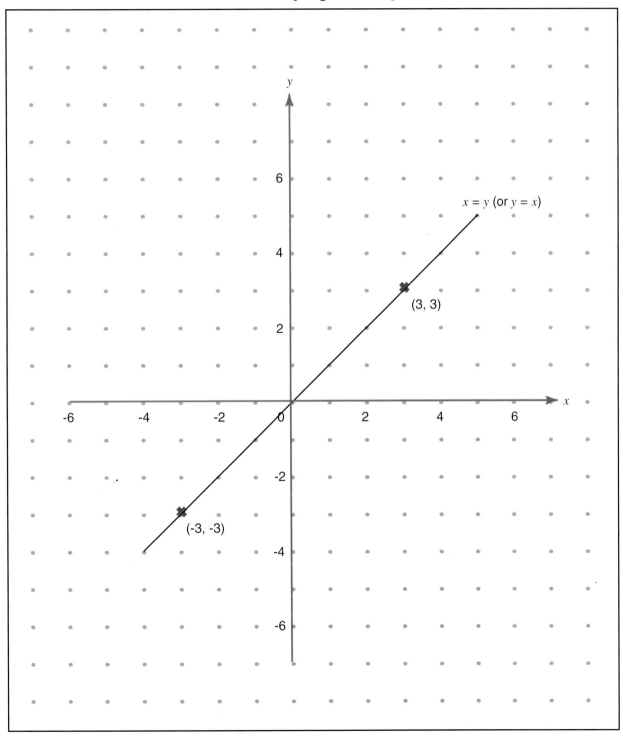

Answers to page 46

2. Table for $y = x + 4$

X	$x + 4 = y$	(x, y)
2	$2 + 4 = 6$	$(2, 6)$
4	$4 + 4 = 8$	$(4, 8)$
6	$6 + 4 = 10$	$(6, 10)$
0	$0 + 4 = 4$	$(0, 4)$
−3	$−3 + 4 = 1$	$(−3, 1)$

54

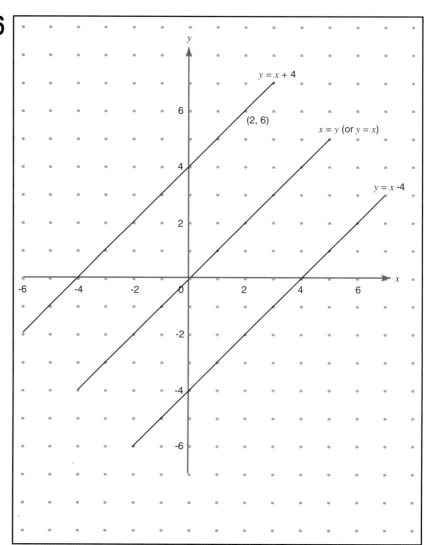

4. Table for $y = x − 4$
You can choose any numbers for x that you like. These are some of the possibilities.

x	$x − 4 = y$	(x, y)
−2	$−2 − 4 = −6$	$(−2, −6)$
0	$0 − 4 = −4$	$(0, −4)$
2	$2 − 4 = −2$	$(2, −2)$
4	$4 − 4 = 0$	$(4, 0)$

Turn back to page 47.

Answer to page 48.

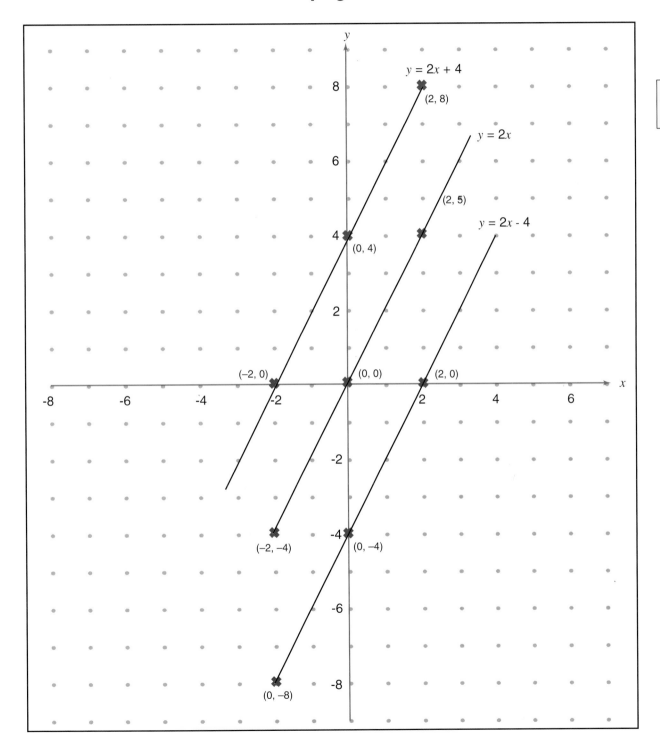

Answers to page 49

1. Table for $y = x^2$

x	$x^2 = y$	(x, y)
−3	$−3^2 = 9$	(−3, 9)
−2	$−2^2 = 4$	(−2, 4)
−1	$−1^2 = 1$	(−1, 1)
0	$0^2 = 0$	(0, 0)
1	$1^2 = 1$	(1, 1)
2	$2^2 = 4$	(2, 4)
3	$3^2 = 9$	(3, 9)

3. Table for $y = x^2 + 2$

x	$x^2 + 2 = y$	(x, y)
−3	$3^2 + 2 =$ 9 + 2 = 11	(−3, 11)
−2	$−2^2 + 2 =$ 4 + 2 = 6	(−2, 6)
−1	$−1^2 + 2 =$ 1 + 2 = 3	(−1, 3)
0	$0^2 + 2 = 2$ 0 + 2 = 2	(0, 2)
1	$1^2 + 2 = 3$	(1, 3)
2	$2^2 + 2 = 6$	(2, 6)
3	$3^2 + 2 = 11$	(3, 11)

Table for $y = x^2 − 3$

x	$x^2 − 3 = y$	(x, y)
−3	9 − 3 = 6	(−3, 6)
−2	4 − 3 = 1	(−2, 1)
−1	1 − 3 = −2	(−1, −2)
0	0 − 3 = −3	(0, −3)
1	1 − 3 = −2	(1, −2)
2	4 − 3 = 1	(2, 1)
3	9 − 3 = 6	(3, 6)

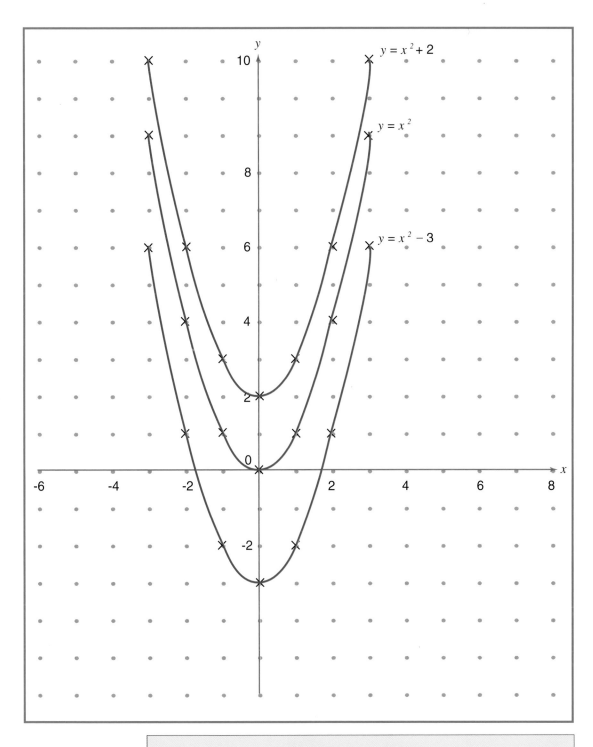

Turn back to page 49.

Table for $y = 3x^2$

x	$3x^2 = y$	(x, y)
−2	$3 \times (−2 \times −2) = 12$	(−2, 12)
1	$3 \times (−1 \times −1) = 3$	(−1, 3)
0	$3 \times 0 = 0$	(0, 0)
1	$3 \times (1 \times 1) = 3$	(1, 3)
2	$3 \times (2 \times 2) = 12$	(2, 12)

Table for $y = 3x^2 + 4$

x	$3x^2 + 4 = y$	(x, y)
−2	$3 \times (−2 \times −2) + 4 = 16$	(−2, 16)
1	$3 \times (−1 \times −1) + 4 = 7$	(−1, 7)
0	$3 \times 0 + 4 = 4$	(0, 4)
1	$3 \times (1 \times 1) + 4 = 7$	(1, 7)
2	$3 \times (2 \times 2) + 4 = 16$	(2, 16)

Table for $y = x^2 − 5$

x	$3x^2 − 5 = y$	(x, y)
−2	$3 \times (−2 \times −2) − 5 = 7$	(−2, 7)
−1	$3 \times (−1 \times −1) − 5 = −2$	(−1, −2)
0	$3 \times 0 − 5 = −5$	(0, −5)
1	$3 \times (1 \times 1) − 5 = −2$	(1, −2)
2	$3 \times (2 \times 2) − 5 = 7$	(2, 7)

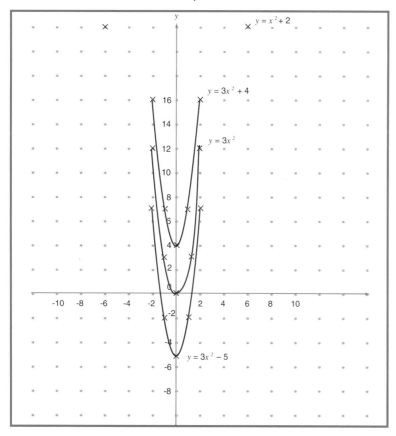

Turn back to page 50.

58

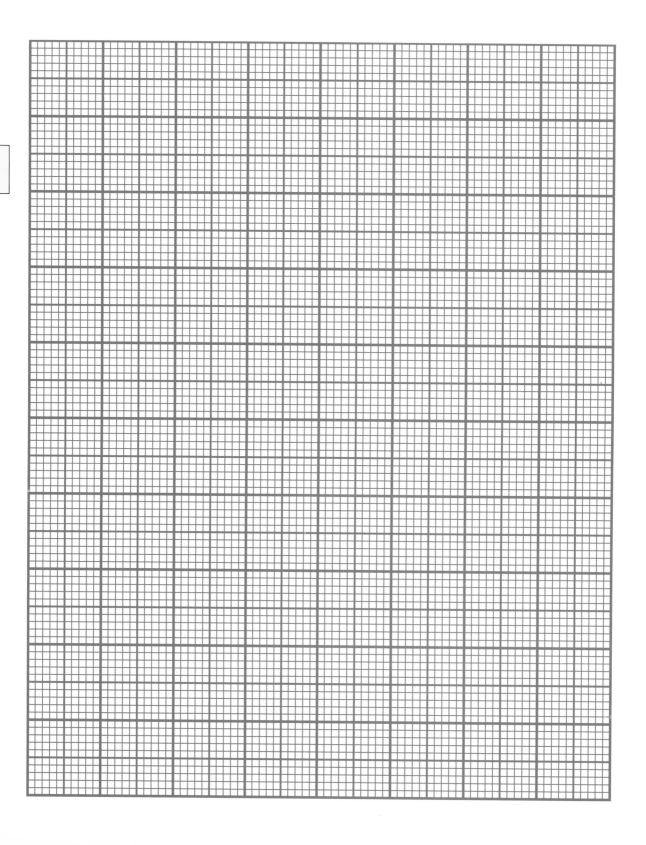

Glossary

axis   a fixed line that is divided into equal parts to allow a point to be given a coordinate. A graph has a horizontal and a vertical axis which meet at right angles at (0, 0).

coordinate   a number that fixes a point in relation to an axis. On a graph, a point can be referred to by a pair of coordinates. The first coordinate tells you how far along the horizontal axis the point lies. The second coordinate tells you how far up or down the vertical axis the point is.

correlation   if there is a connection between two sets of measurements they are said to have a correlation. For example, there is some correlation between gender and speed of tennis serve. Most men will serve faster then most women, although some women can serve faster than some men. There is probably no correlation, in adults, between height and the ability to draw cartoons.

equation   a statement that two things are equal. In mathematics, letters are often used to stand for the things. $x = y - 4$ could mean "the amount in dollars of pocket money you get is the same as your age minus 4." If you are ten years old, you get six dollars (10 – 4). The line on a graph is made up of all the points for which a particular equation is true.

horizontal   a line drawn from side to side

legionary   a soldier of ancient Rome

negative numbers   a – sign in front of a number tells you the number is less than zero. A temperature of –10° is 10° below zero. – 7 means 7 less than 0. You can put a + sign in front of numbers greater than 0, but we do not usually bother.

parallel   lines that are parallel always remain the same distance apart. Railway tracks are made from parallel rails.

vertical   a line drawn from top to bottom

Answers

Page 9
1. Tony's grandmother owes his dad $14. (His dad has to run 4 km to earn $1. (56 ÷ 4 = 14)
2. 70 km = 44 miles approximately.

Page 10
1. The graph line is parallel to the horizontal axis.

Page 11
Story **A** matches graph **2**.
Story **B** matches graph **1**.
Graph **3** shows a distance being traveled in no time, which is impossible.
Part of graph **4** shows a journey back in time, which is impossible.

Page 12
Graph **A** matches the drawing of the boy, it shows that he lost weight during a period of weeks.
Graph **C** matches the drawing of the tree from being a seed through its growth, and finally to when it was chopped down. Graph **B** matches the rose bush drawing, showing its growth, then it being pruned, and then it growing again.

Page 13
See page 14.

Page 15
Graph **B** shows Tammy taking a bath.

Page 17
The ball is thrown from someone's hand. It does not start from the ground.

Page 19
See page 20.

Page 21
1. 1 small square on the vertical axis represents 1° Fahrenheit.

Page 22
1. 5 small squares = 24 hours.
2. 3 small squares = 1° F.

Page 24
1. Whizzo gives you more detergent. Your graph should look like the one on page 26.

Page 28
1. 3½"

Page 29
1. A curve
2. (a) approximately 8.5 hours (answers from 8.3 to 8.7 hours are acceptable).
(b) approximately 6.9 hours (answers from 6.7 to 7.1 hours are acceptable).
3. The journey will take about 11 hours, so you have to leave about 1 a.m.

Page 30
Aaron is standing to the left of Gerald, with Hanah and then Fran to the right.

Page 31
1. Patsy will be about 56" tall.

Page 32
1. Each cross tells you the height and the number of pets of a particular person.
2. Gerald has no pets.

Page 39
1. Ruth threw a 2 on the green die.
2. No, Bob threw

Page 40
1. Bob
2. (1, 5)
3. The dice only go up to 5.
4. Your hits should all be on the line, although some of the crosses may be missing.

Page 41
1. There are 5 possible throws to score a hit.
2. Your hits should be on the same line.
3. (0, 0), (1, 1), (2, 2), (3, 3), (4, 4), (5, 5)

4. r + g = 6

5.

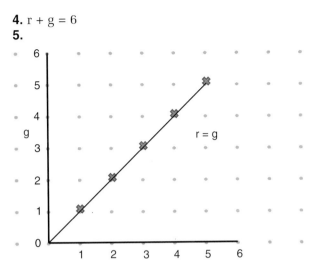

6. (0, 0), (1, 1), (2, 2), (3, 3), (4, 4), (5, 5)

7. Yes, they are the same.

Page 42

1. There are 4 throws that can score a hit.

2. (5, 2), (4, 3), (3, 4), (2, 5).

3. r + g = 7.

4. Each player might score two hits.

Page 43

A (3, 1), **B** (1, −2), **C** (−4, −2), **D** (−2, 5)

Page 44

1. A square.

2. E (2, 2), **F** (2, −2), **G** (−2, −2), **H** (−2, 2).

3. (3, 1) and (1, 3).   **4.** $x + y = 4$.

Page 45

1. D (1.5, −2.5) or (1½, −2½),

E (-0.75, 1) or (−¾, 1).

Answers to Coordinates Check Up, see page 52.

Page 47

1. The line $y = x - 4$ cuts the y axis at − 4 and the x axis at 4.

2. The line $y = x + 2$ will be parallel to the others. It will cut the y axis at 2 and the x axis at −2.

3. All the equations of the form $y = x +$ or $-$ any number will be parallel to each other.

Page 48

1. $y = 2x$

x	$2x = y$	(x, y)
−2	2 × − 2 = −4	(−2, −4)
0	2 × 0 = 0	(0, 0)
2	2 × 2 = 4	(2, 4)

$y = 2x + 4$

x	$2x + 4 = y$	(x, y)
−2	2 × −2 + 4 = 0	(−2, 0)
0	2 × 0 + 4 = 4	(0, 4)
2	2 × 2 + 4 = 8	(2, 8)

$y = 2x - 4$

x	$2x - 4 = y$	(x, y)
−2	2 × −2 − 4 = −8	(−2, −8)
0	2 × 0 − 4 = −4	(0, −4)
2	2 × 2 − 4 = 0	(2, 0)

2. See page 55.

3. Yes, they are parallel.

4. $y = 2x$ cuts the y and x axes at (0, 0);
$y = 2x + 4$ cuts the y axis at 4 and the x axis at −2;
$y = 2x - 4$ cuts the y axis at −4 and the x axis at 2.

5. Each is a set of parallel lines.

6. The second set has a steeper slope.

7. $y = 5x$ is not parallel to the other three lines on the page.

Page 49

See pages 56 and 57 for the tables and graphs.

1. The curves are all the same but in different positions on the y axis.

2. The curve is symmetrical. The tables and curves for the second group of equations are shown on page 58.

3. Yes.

4. No.

5. It changes the shape of the curve.

Page 51

1. (a) 5, **(b)** ½, **(c)**1. (see page 55 for graph).

Index